AF782534

CONTENTS

MAY 1, 2007

INTERNET SPYWARE (I-SPY) PREVENTION ACT OF 2007, AND THE SECURING AIRCRAFT COCKPITS AGAINST LASERS ACT OF 2007

HEARING

BEFORE THE

SUBCOMMITTEE ON CRIME, TERRORISM, AND HOMELAND SECURITY

OF THE

COMMITTEE ON THE JUDICIARY
HOUSE OF REPRESENTATIVES

ONE HUNDRED TENTH CONGRESS

FIRST SESSION

ON

H.R. 1525 and H.R. 1615

MAY 1, 2007

Serial No. 110–109

Printed for the use of the Committee on the Judiciary

Available via the World Wide Web: http://judiciary.house.gov

U.S. GOVERNMENT PRINTING OFFICE

35–113 PDF WASHINGTON : 2007

For sale by the Superintendent of Documents, U.S. Government Printing Office
Internet: bookstore.gpo.gov Phone: toll free (866) 512–1800; DC area (202) 512–1800
Fax: (202) 512–2104 Mail: Stop IDCC, Washington, DC 20402–0001

COMMITTEE ON THE JUDICIARY

JOHN CONYERS, Jr., Michigan, *Chairman*

HOWARD L. BERMAN, California
RICK BOUCHER, Virginia
JERROLD NADLER, New York
ROBERT C. "BOBBY" SCOTT, Virginia
MELVIN L. WATT, North Carolina
ZOE LOFGREN, California
SHEILA JACKSON LEE, Texas
MAXINE WATERS, California
MARTIN T. MEEHAN, Massachusetts
WILLIAM D. DELAHUNT, Massachusetts
ROBERT WEXLER, Florida
LINDA T. SÁNCHEZ, California
STEVE COHEN, Tennessee
HANK JOHNSON, Georgia
LUIS V. GUTIERREZ, Illinois
BRAD SHERMAN, California
TAMMY BALDWIN, Wisconsin
ANTHONY D. WEINER, New York
ADAM B. SCHIFF, California
ARTUR DAVIS, Alabama
DEBBIE WASSERMAN SCHULTZ, Florida
KEITH ELLISON, Minnesota

LAMAR SMITH, Texas
F. JAMES SENSENBRENNER, Jr., Wisconsin
HOWARD COBLE, North Carolina
ELTON GALLEGLY, California
BOB GOODLATTE, Virginia
STEVE CHABOT, Ohio
DANIEL E. LUNGREN, California
CHRIS CANNON, Utah
RIC KELLER, Florida
DARRELL ISSA, California
MIKE PENCE, Indiana
J. RANDY FORBES, Virginia
STEVE KING, Iowa
TOM FEENEY, Florida
TRENT FRANKS, Arizona
LOUIE GOHMERT, Texas
JIM JORDAN, Ohio

Perry Apelbaum, *Staff Director and Chief Counsel*
Joseph Gibson, *Minority Chief Counsel*

Subcommittee on Crime, Terrorism, and Homeland Security

ROBERT C. "BOBBY" SCOTT, Virginia, *Chairman*

MAXINE WATERS, California
WILLIAM D. DELAHUNT, Massachusetts
JERROLD NADLER, New York
HANK JOHNSON, Georgia
ANTHONY D. WEINER, New York
SHEILA JACKSON LEE, Texas
MARTIN T. MEEHAN, Massachusetts
ARTUR DAVIS, Alabama
TAMMY BALDWIN, Wisconsin

J. RANDY FORBES, Virginia
LOUIE GOHMERT, Texas
F. JAMES SENSENBRENNER, Jr., Wisconsin
HOWARD COBLE, North Carolina
STEVE CHABOT, Ohio
DANIEL E. LUNGREN, California

Bobby Vassar, *Chief Counsel*
Michael Volkov, *Minority Counsel*

INTERNET SPYWARE (I-SPY) PREVENTION ACT OF 2007, AND THE SECURING AIRCRAFT COCKPITS AGAINST LASERS ACT OF 2007

TUESDAY, MAY 1, 2007

HOUSE OF REPRESENTATIVES,
SUBCOMMITTEE ON CRIME, TERRORISM,
AND HOMELAND SECURITY
COMMITTEE ON THE JUDICIARY,
Washington, DC.

The Subcommittee met, pursuant to notice, at 1 p.m., in Room 2141, Rayburn House Office Building, the Honorable Robert Scott (Chairman of the Subcommittee) presiding.

Present: Representatives Scott, Waters, Johnson, Forbes, Gohmert, Coble, Chabot, and Lungren.

Staff present: Bobby Vassar, Subcommittee Chief Counsel; Ameer Gopalani, Majority Counsel; Veronica Eligan, Professional Staff Member; Caroline Lynch, Minority Counsel; and Kelsey Whitlock, Minority Staff Assistant.

Mr. SCOTT. The Subcommittee will now come to order.

I am pleased to welcome you to today's hearing before the Subcommittee on Crime, Terrorism, and Homeland Security.

The first bill we will consider will be H.R. 1525, the "Internet Spyware (I-SPY) Prevention Act of 2007." And the other bill is H.R. 1615, the "Securing Aircraft Cockpits Against Lasers Act of 2007."

The first bill we consider, H.R. 1525, amends Title 18 of the U.S. Code to impose criminal penalties on those who use spyware to perpetrate identity theft and numerous other privacy intrusions on innocent Internet users. It provides resources and guidance to the Department of Justice for prosecuting these offenses.

The House passed similar legislation in the 109th Congress, and we will be hearing from the chief sponsors of the prior and current legislation, Congresswoman Zoe Lofgren of California and Congressman Bob Goodlatte of Virginia. I would like to thank and commend them for developing and moving these bills on a bipartisan basis.

I will introduce my entire statement in its entirety, but move on to H.R. 1615, Securing Aircraft Cockpits Against Lasers Act of 2007.

I would like to welcome our colleague, Congressman Ric Keller, who is a Member of the Judiciary Committee, as are Bob Goodlatte and Zoe Lofgren. Congressman Keller has been instrumental in bringing attention to the issue of the danger of lasers composed on aircraft, and I look forward to his testimony.

He introduced this bill in the 109th Congress, and I joined them in cosponsoring that bill and continue to support the legislation now.

The purpose of this bill is to address the problems of individuals aiming lasers at cockpits of aircraft, particularly at the critical stages of takeoff and landing. This practice constitutes a threat to aviation security and passenger safety. It adds a section following 18 USC Section 38 to impose criminal penalties upon any individual who knowingly aims their laser pointer at an aircraft within the special aircraft jurisdiction of the United States.

It includes fines of up to $250,000 and imprisonment of up to 5 years. And, again, I will introduce the entirety of my statement on that bill into the record.

At this point I will call on my colleague from Virginia, the Ranking Member of the Subcommittee, Mr. Forbes, for his statement.

Mr. FORBES. Thank you, Mr. Chairman. I will follow your lead. I know we have three very busy Members who are here to testify today. So with your permission, I will put the entirety of my statement in the record.

But I do just want to thank you for holding this hearing and also thank Congresswoman Lofgren and Congressman Goodlatte for the work that they have done on H.R. 1525 and certainly recognize Congressman Keller's commitment to aircraft safety and the work that he has done on 1615.

I am proud to be a cosponsor of both these bills we are considering today, and I urge my colleagues to support them.

[The prepared statement of Mr. Forbes follows:]

PREPARED STATEMENT OF THE HONORABLE J. RANDY FORBES, A REPRESENTATIVE IN CONGRESS FROM THE STATE OF VIRGINIA, AND RANKING MEMBER, SUBCOMMITTEE ON CRIME, TERRORISM, AND HOMELAND SECURITY

Statement of Ranking Member Randy Forbes
Subcommittee on Crime, Terrorism and Homeland Security

Legislative Hearing on H.R. 1525, "The Internet Spyware (I-Spy) Prevention Act of 2007," and H.R. 1615, "Securing Aircraft Cockpits Against Lasers Act of 2007"

May 1, 2007

Thank you, Chairman Scott. I appreciate your holding this legislative hearing on H.R. ~~1592,~~ 1525 "The Internet Spyware (I-Spy) Prevention Act of 2007," and H.R. 1615, "Securing Aircraft Cockpits Against Lasers Act of 2007."

I want to recognize my colleagues, Congresswoman Lofgren and Congressman Goodlatte, who are here to testify in support of H.R. 1525, and thank them for their dedication on this important issue. I also want to recognize my colleague, Congressman Keller, who is here to testify in support of H.R. 1615, and thank him for his commitment to aircraft safety.

Spyware is a growing and serious problem. Spyware is software that allows criminals to crack into computers to alter a user's security settings, collect personal information to steal a user's identity, or to commit other crimes.

H.R. 1525 is bipartisan legislation which imposes criminal penalties on computer hacking intrusions and the use of spyware. A maximum term of 5 years imprisonment can be imposed for a hacking violation in which an unauthorized user accesses a computer. In addition, a maximum term of 2 years imprisonment can be imposed for anyone who uses spyware to break into a computer and alter the security settings or obtain personal information about a person.

The bill also authorizes $10 million for fiscal years 2008 to 2011 for the Department of Justice to increase federal prosecutions of these new offenses.

H.R. 1615, the "Securing Aircraft Cockpits Against Lasers Act of 2007, amends the Federal criminal code to prohibit aiming a laser pointer at an aircraft or at the flight of an aircraft in the special aircraft jurisdiction of the United States.

Over the past several years, there have been an increasing number of reports to the Federal Aviation Administration of the aiming of lasers into airplane cockpits. The FAA reports over 400 incidents since 1990. FAA research has shown that laser illuminations can temporarily disorient or disable a pilot during critical stages of flight such as landing or take-off. They can even cause temporary blindness. In some cases, these laser illuminations cause permanent damage. This type of interference, whether it is in an intentional effort to sabotage a plane, or just a prank, should not be tolerated because of the potential for catastrophe. This bill is a good, common sense measure aimed at deterring and prosecuting those that commit a senseless act of potential sabotage.

I am proud to be a cosponsor of both bills we are considering today and I urge my colleagues to support them.

Mr. SCOTT. Thank you.

I will now introduce our witnesses.

The first witness will be the Honorable Zoe Lofgren, who has represented California's 16th Congressional District since 1994. She currently serves as Chair of the Judiciary Committee Subcommittee on Immigration, Citizenship, Refugees, Border Security, and International Law. She also serves on the Homeland Security and House Administration Committees and serves as Chair of the California delegation Democratic Congressional Delegation. She has headed a number of technology initiatives, including the E-Rate, which provides affordable Internet access to schools, libraries and rural health centers. She has a B.A. from Stanford and a J.D. from the University of Santa Clara School of Law.

Our next witness will be Congressman Bob Goodlatte. Congressman Goodlatte began his eighth term representing the 6th Congressional District of Virginia in 2007. He is co-chair of the bipartisan Congressional International Caucus, Chairman of the House Republican Higher Technology Working Group and co-chair of the Congressional International Anti-Piracy Caucus. He serves on the Judiciary Committee and is the ranking Republican on the Agriculture Committee. He is a graduate of Washington Lee University School of Law and has an undergraduate degree in Government from Bates College in Maine.

Our final witness will be Congressman Ric Keller. He is the Ranking Member of the House Committee on Education and Labor's Higher Education Subcommittee. He was re-elected in November 2006 to his fourth term. He represents the 8th District in Florida, which covers the greater Orlando area, where he grew up. He received his Bachelor's degree from East Tennessee State University and his law degree from Vanderbilt. And I have the privilege of serving on two Committees with Mr. Keller. We both serve on the Education and Labor Committee as well as the Judiciary Committee.

Each of the witnesses' written statements will be made a part of the record in their entirety, and I would ask each of our witnesses to summarize his or her testimony in 5 minutes or less.

I would recognize Mr. Gohmert as being present. If he has a statement, he was allowed to enter it into the record. But I would like to recognize his presence.

At this time, Congresswoman Lofgren.

TESTIMONY OF THE HONORABLE ZOE LOFGREN, A REPRESENTATIVE IN CONGRESS FROM THE STATE OF CALIFORNIA

Ms. LOFGREN. Thank you, Mr. Chairman and Ranking Member Forbes.

I am very proud to have partnered with my colleague, Bob Goodlatte, on this legislation to combat spyware, as we have in previous Congresses.

Spyware is actually becoming one of the biggest threats to consumers on the Internet, and it is one of the reasons why we have such an identity theft epidemic. Thieves use spyware to harvest personal information from unsuspecting Americans and they even use spyware to track keystrokes an individual makes, including credit card and Social Security numbers.

Spyware can also adversely affect the business community who have to spend money to block and remove it from their systems. Microsoft, in fact, has stated that spyware is "at least partially responsible for approximately one-half of all application crashes reported to them."

Experts estimate that as many as 80 percent to 90 percent of all personal computers contain some form of spyware. In 2004, 93 million Americans experienced a spyware-related problem. Consumers spent $2.6 billion last year trying to block spyware or remove it from their system.

In short, spyware is a real problem that is endangering consumers, damaging businesses and creating millions of dollars in additional cost.

H.R. 1525 is a bipartisan measure that identifies the truly unscrupulous acts associated with spyware and subjects them to criminal punishment. The bill is important because it focuses on behavior, not technology, and it targets the worst form of spyware without unduly burdening technological innovation.

The bill also funds the attorney general to find and prosecute spyware offenders and phishing scam artists and it expresses the sense of Congress that the Department of Justice should pursue online phishing scams where criminals send fake e-mails to consumers on behalf of well-known companies.

Phishing and spyware aren't just inconvenient to consumers. They represent a threat to the vitality of the Internet. If you can't trust the Internet, people will not use the Internet for commerce, and that is not a good thing.

Focusing on bad actors and criminal conduct is very much preferable to an approach that criminalizes technology or imposes notice of consent requirements. Bad actors don't comply with requirements and I think the Can-Spam Act of a few Congresses ago is evidence of the futility of pursuing that approach.

The more notices Internet users receive, the less likely they are to pay attention to any of them; 75 percent of users don't read agreements, privacy statements or disclaimers on the Internet. And in 2005, the Pew Internet and American Life Project proved this point. A diagnostic site included a clause in one of its agreements that promised $1,000 to the first person to write in and request the money. The agreement was downloaded more than 3,000 times before someone finally claimed the reward.

We don't want to over-regulate the user experience. We must avoid interfering with the increasingly seamless, intuitive and interactive online environment. Regulation of technology is almost always a bad idea because technology changes faster than Congress can legislate and what we attempt to regulate will morph into something else and render useless the regulatory scheme that we adapt.

Legislation that attempts to control technology can also have the pernicious effect of chilling innovation by chilling investment by venture capital sectors into prohibited technological arenas. 1525 avoids these pitfalls by focusing on bad conduct. It does not prevent existing or future State laws that prohibit spyware. The bill preempts only civil actions that are based on violations of this new Federal criminal law. It does not prevent a State from passing a similar law nor does it prevent any lawsuits that are premised on existing State laws.

I am honored that this bill has strong support from some of the biggest names in technology, including Microsoft and Dell. It is also supported by the U.S. Chamber of Commerce, the Center for Democracy and Technology and even the Distributing Computing Industry Association.

The bill has had broad support in past Congresses. In the last Congress, the floor vote was 395 to 1. So what we are doing here today is important for consumers, for businesses and for the future of our high tech economy.

I am grateful to my colleague, Mr. Goodlatte, for his leadership in this Congress and in past Congresses.

And I note, as we begin the legislative process, we are certainly very open to any improvement or tweaking that might be necessary, but we also think this is a very solid effort.

And I thank the Committee for your attention and yield back.

[The prepared statement of Ms. Lofgren follows:]

PREPARED STATEMENT OF THE HONORABLE ZOE LOFGREN, A REPRESENTATIVE IN CONGRESS FROM THE STATE OF CALIFORNIA

Chairman Scott, Ranking Member Forbes, and distinguished members of the subcommittee, thank you for inviting me to speak before you today on the growing threat to Internet users and Internet commerce posed by spyware and phishing scams, and on the way that the Internet Spyware (I-SPY) Prevention Act of 2007 will counter that threat.

Spyware is a serious and growing problem for American consumers and businesses. Thieves are using spyware to harvest personal information such as Social Security numbers and credit card numbers for use in a variety of criminal enterprises. Although the definition of spyware is a moving target, the FTC loosely defines the term as software that "aids in gathering information about a person or organization without their knowledge and which may send such information to another entity without the consumer's consent, or asserts control over a computer without the consumer's knowledge." The Anti-Spyware Coalition offers a slightly different definition of spyware as "technologies deployed without the appropriate user consent and/or implemented in ways that impair user control," including:

- Material changes that affect user experience, privacy, or system security;
- Use of system resources, including what programs are installed on computers; and/or
- Collection, use, and distribution of personal or other sensitive information.

Two of the most serious forms of spyware are "keystroke loggers" that capture every key typed on a particular computer, allowing cyber-criminals to gain access to credit card accounts and other personal information, and programs that hijack users' system settings.

Nine out of every ten Internet users have modified their online behavior out of fear of falling victim to spyware. Indeed, consumers spent $2.6 billion last year trying to block or remove spyware from their computers. But consumers are seldom successful at completely eliminating spyware from their systems. Recent studies estimate that 80 percent of computers are infected with some form of spyware and 89 percent of consumers are unaware of that fact. 93 million American adults experienced a spyware-related problem in 2004. As broadband reaches American communities that have less experience with the online world, the number of victims of spyware will almost certainly increase.

Spyware is as much a problem for technology companies and other businesses as it is for individuals. Microsoft analysts have reported that spyware is at least partially responsible for about one-half of all the application crashes that are reported to them. Spyware is also threat to the Internet as a whole. Just this February, a massive denial-of-service attack targeted DNS root servers, including one maintained by the Department of Defense. Although the source of the ultimately unsuccessful attack was unclear, hijacked computers are often turned into "zombies" that participate in denial-of-service attacks without the knowledge of their users.

As the Judiciary committee has noted in the past, there is no "silver bullet" for ending spyware. Instead, we must rely on a multi-pronged approach that involves greater consumer awareness, the use of available technological countermeasures, and an effective criminal enforcement strategy. The legislation you are considering today is a crucial component of this last prong. That is why I was pleased to work once again with Representative Goodlatte to introduce the I-SPY Prevention Act.

The Act imposes significant criminal penalties for the most serious and prevalent criminal activities that employ spyware. For example, the Act would impose a prison sentence of up to 5 years for use of spyware in furtherance of another Federal crime. The Act also imposes up to a 2-year sentence for hacking into a computer and altering its security settings or obtaining personal information with the intent to defraud or injure the person or damage a computer.

The Act also assists the Department of Justice in enforcing these new provisions. The legislation authorizes $10 million in funding for fiscal years 2008 through 2011 for prosecutions to deter the use of spyware as well as "phishing" scams. Phishing scams involve criminals using websites or e-mail addresses that mimic those of well-known and legitimate businesses to deceive Internet users into revealing personal information that can be used to defraud them.

The central feature of the Act is that it targets bad actors and bad behavior without unduly restricting innovation in the online universe. As the Judiciary committee and other entities have noted, one of the greatest difficulties in solving the spyware problem is that many legitimate and beneficial tools for making a user's Internet experience more enjoyable and productive are technologically indistinguishable from spyware that is used to harm users and computers. For example, an Internet "cookie" can be used to store detailed information about a user's preferences when visiting a much-frequented website. But the same technology can be used by identity thieves to track and store personal and financial information. The appropriate legislative target is not the cookie itself, but the criminals who use it for illegal purposes. The I-SPY Prevention Act is a measured and careful approach to combating spyware that captures this distinction.

Other legislative approaches revolve around notice-and-consent procedures that require computer users to be notified and either "opt in" or "opt out" of installing code at the time of installation. Ensuring user consent is critical, as is implicit in the term "authorized access" contained in the I-SPY Act and in existing Section 1030. In my view, however, a notice-and-consent approach is ill-advised for three reasons.

First, bad actors—the criminals we should be most concerned about—are unlikely to comply with that requirement. As we learned with the CAN-SPAM Act, legislatively mandating a certain approach is a far cry from ensuring that others comply with it. Thus, legitimate uses of technology will be burdened by notice-and-consent requirements while bad actors will most likely ignore them.

Second, the more notices and warnings that Internet users see, the less likely they are to pay attention to any single one. In 2005, the Pew Internet & American Life Project proved this point. A diagnostic site included a clause in one of its user agreements that promised $1,000 to the first person to write in and request the money. The agreement was downloaded more than 3,000 times before someone finally read the fine print and claimed the reward. Additionally, a Pew survey found that 73 percent of Internet users said that they do not always read user agreements, privacy statements, or other disclaimers before downloading or installing programs.

Finally, and most importantly, we must take care not to legislate the online user experience. Internet users have come to expect and demand a seamless, intuitive, and interactive experience with their online environment. Those expectations have led to the development of social networking and bookmarking sites, "wikis," and an explosion in user-generated content. Users are interacting with the Internet in a way that allows them to shape and control their online experience to a degree that, until recently, would have been unimaginable. This has been a tremendous boon to both consumers and the American economy. It would be unwise and unfortunate if we were to interfere with the continued evolution of the Internet through overbroad regulation.

The I-SPY Prevention Act avoids these pitfalls by focusing attention and resources where they are needed most, on criminal enterprises that harm Internet users and Internet commerce. That is why the Act also expresses the sense of Congress that the Department of Justice should use the Act to prosecute vigorously those who use spyware to commit crimes and those that conduct phishing scams.

Finally, I wish to clarify the Act's provision addressing state civil actions. Some people have construed § 1030A(c) as a bar on any civil action premised on conduct that violates the Act. That construction is incorrect. The Act merely states that violation of the Act itself cannot supply the basis for a state civil action. This provision is necessary because some States permit tort claims based on the violation of Federal criminal statutes. Were we to allow the Act to serve as the basis for tort claims in multiple jurisdictions, we would wind up with multiple and inconsistent state-court interpretations of the Act. Because much of the power and promise of the Internet comes from its ability to transcend geographic and political boundaries, we must avoid miring Internet commerce in potentially inconsistent state applications of Federal law. Section 1030A(c) ensures that this does not happen. At the same time, that provision does not preempt state-court cases based on independent state-law causes of action. Nor does it preempt actions of any kind in Federal court.

In closing, I simply note that a very broad coalition of high technology industries, commercial organizations, and public interest groups have come together to support this legislation. The breadth of the support for this bill extends to the House itself. When Representative Goodlatte and I brought this legislation to the floor in the past two Congresses it passed by an overwhelming majority. Indeed, the floor vote in the 109th Congress was 395–1. That support was there for a reason. Spyware is a serious and growing problem and the I-SPY Prevention Act is the right way to fight it.

I applaud the subcommittee for once again focusing on this very important piece of legislation. Thank you for the opportunity to testify today.

Mr. SCOTT. Mr. Goodlatte?

TESTIMONY OF THE HONORABLE BOB GOODLATTE, A REPRESENTATIVE IN CONGRESS FROM THE STATE OF VIRIGINIA

Mr. GOODLATTE. Thank you, Chairman Scott and Ranking Member Forbes and the other Members of the Subcommittee for allowing me to testify at this important hearing. My full written testimony has been submitted for the record.

And I am very pleased to join with my colleague from California, representative Zoe Lofgren, with whom I have worked on many Internet-and technology-related issues, in the reintroduction of H.R. 1525, the Internet Spyware, or I-SPY, Prevention Act.

This bipartisan legislation will impose tough criminal penalties on those that use software for nefarious purposes without imposing a broad regulatory regime on legitimate online businesses. I believe that this targeted approach is the best way to combat spyware.

The continued growth of the Internet has brought tremendous enhancements to our quality of life, from advances in the delivery of healthcare to the ability of consumers to instantaneously conduct transactions online. Increasingly, consumers want a fast connection to the Internet and want the delivery of online services to be seamless and online service providers have invested significant resources to develop software to make their services as safe, reliable, and fast as possible.

However, the Internet will never reach its full potential until consumers feel safe to conduct transactions online. One enormous hurdle to consumer confidence in the Internet is the purveyance of spyware. Unfortunately, similar types of software to what legitimate businesses use to deliver new and innovative services can also be used by bad actors to break into computers, steal personal information and commit identity theft and other crimes.

The term “spyware” is used to describe software that criminals use to secretly crack into computers to conduct nefarious activities such as altering a user’s security settings, collecting personal information to steal a user’s identity or committing other crimes.

A recent study done by the National Cybersecurity Alliance revealed that over 90 percent of consumers had some form of spyware on their computers and most consumers were not aware of it. With the interstate nature of the Internet, Congress clearly has a role to play in both punishing those that use software to commit online crimes and preventing the continuing erosion of consumer confidence in the Internet.

However, as Congress considers legislation in this area, I believe that four overarching principles should guide the development of any software legislation. First, we must punish the bad actors while protecting legitimate online companies. Second, we must not over-regulate but rather encourage innovative new services and the growth of the Internet. Third, we must not stifle the free market interactions between consumers and service providers. And, fourth, we must target the behavior, not the technology.

The I-SPY Prevention Act would impose criminal penalties on the most egregious behaviors associated with spyware. Specifically, this legislation would impose up to a 5-year prison sentence on anyone who uses spyware to intentionally break into a computer and uses that software in furtherance of another Federal crime.

In addition, it would impose up to a 2-year prison sentence on anyone who uses spyware to intentionally break into a computer, and either alter the computer's security settings or obtain personal information with the intent to defraud or injure a person or with the intent to damage a computer.

By imposing stiff penalties on these bad actors, this legislation will help deter the use of spyware and will thus help protect consumers from these aggressive attacks.

Enforcement is crucial in combating spyware. The I-SPY Prevention Act authorizes $10 million for fiscal years 2008 through 2011 to be devoted to prosecutions involving spyware, phishing and pharming scams.

Phishing scams occur when criminals send fake e-mail messages to consumers on behalf of famous companies and request account information that is later used to conduct criminal activities.

Pharming, an even more nefarious practice, occurs when hackers redirect Internet traffic to fake sites in order to steal personal information, such as credit card numbers, passwords and account information.

In summary, this I-SPY Prevention Act is a targeted approach that protects consumers by imposing stiff penalties on the truly bad actors while protecting the ability of legitimate online companies to develop new and exciting products and services for consumers.

Mr. Chairman, thank you again for the opportunity to testify before the Subcommittee.

[The prepared statement of Mr. Goodlatte follows:]

PREPARED STATEMENT OF THE HONORABLE BOB GOODLATTE, A REPRESENTATIVE IN CONGRESS FROM THE STATE OF VIRGINIA

Chairman Scott, Ranking Member Forbes, and members of the Subcommittee, thank you for inviting me to testify at this important hearing.

I was pleased to join with my colleague from California, Representative Zoe Lofgren, to reintroduce H.R. 1525, the "Internet Spyware (I-SPY) Prevention Act." This bi-partisan legislation will impose tough criminal penalties on those that use software for nefarious purposes, without imposing a broad regulatory regime on legitimate online businesses. I believe that this targeted approach is the best way to combat spyware.

The continued growth of the Internet has brought tremendous enhancements to our quality of life—from advances in the delivery of health care, to the ability of consumers to seamlessly and instantaneously conduct transactions online. Increasingly, consumers want a fast connection to the Internet and want the delivery of online services to be seamless, and online service providers have invested significant resources to develop software to make their services as safe, reliable and fast as possible.

However, the Internet will never reach its full potential until consumers feel safe to conduct transactions online. One enormous hurdle to consumer confidence in the Internet is the purveyance of spyware. Unfortunately, similar types of software to what legitimate businesses use to deliver new and innovative services can also be used by bad actors to break into computers, steal personal information and commit identity theft and other crimes.

Spyware is software that provides a tool for criminals to secretly crack into computers to conduct nefarious activities, such as altering a user's security settings, collecting personal information to steal a user's identity, or to commit other crimes. A recent study done by the National CyberSecurity Alliance revealed that over 90% of consumers had some form of spyware on their computers and most consumers were not aware of it. With the interstate nature of the Internet, Congress clearly has a role to play in punishing those that use software to commit online crimes and thus prevent the continuing erosion of consumer confidence in the Internet.

However, as Congress considers legislation in this area I believe that four overarching principles should guide the development of any spyware legislation. First, we must punish the bad actors, while protecting legitimate online companies. Second, we must not over-regulate, but rather encourage innovative new services and the growth of the Internet. Third, we must not stifle the free market. Fourth, we must target the behavior, not the technology.

The I-SPY Prevention Act would impose criminal penalties on the most egregious behaviors associated with spyware. Specifically, this legislation would impose up to a five-year prison sentence on anyone who uses software to intentionally break into a computer and uses that software in furtherance of another federal crime. In addition, it would impose up to a two year prison sentence on anyone who uses spyware to intentionally break into a computer and either alter the computer's security settings or obtain personal information with the intent to defraud or injure a person or with the intent to damage a computer. By imposing stiff penalties on these bad actors, this legislation will help deter the use of spyware, and will thus help protect consumers from these aggressive attacks.

Enforcement is also crucial in combating spyware. The I-SPY Prevention Act authorizes $10 million for fiscal years 2008 through 2011, to be devoted to prosecutions involving spyware, phishing and pharming scams, and expresses the sense of Congress that the Department of Justice should vigorously enforce the laws against these crimes. Phishing scams occur when criminals send fake e-mail messages to consumers on behalf of famous companies and request account information that is later used to conduct criminal activities. Pharming scams occur when hackers redirect Internet traffic to fake sites in order to steal personal information such as credit card numbers, passwords and account information. This form of online fraud is particularly egregious because it is not as easily discernable by consumers. With pharming scams, innocent Internet users simply type the domain name into their web browsers, and the signal is re-routed to the devious website.

The I-SPY Prevention Act is a targeted approach that protects consumers by imposing stiff penalties on the truly bad actors, while protecting the ability of legitimate companies to develop new and exciting products and services online for consumers.

The I-SPY Prevention Act also avoids excessive regulation and its repercussions, including the increased likelihood that an overly regulatory approach focusing on technology would have unintended consequences that could discourage both consumer use of the Internet as well as the creation of new and exciting technologies and services on the Internet. By encouraging innovation, the I-SPY Prevention Act will help ensure that consumers have access to cutting-edge products and services at lower prices.

In addition, the approach of the I-SPY Prevention Act does not interfere with the free market principle that a business should be free to react to consumer demand by providing consumers with easy access to the Internet's wealth of information and convenience. Increasingly, consumers want a seamless interaction with the Internet, and we must be careful to not interfere with businesses' abilities to respond to this consumer demand with innovative services. The I-SPY Prevention Act will help ensure that consumers, not the federal government, define what their interaction with the Internet looks like.

Finally, by going after the criminal behavior associated with the use of spyware, the I-SPY Prevention Act recognizes that not all software is spyware and that the crime does not lie in the technology itself, but rather in actually using the technology for nefarious purposes. People commit crimes, not software.

Thank you again for the opportunity to testify before the Subcommittee. I look forward to answering any questions you may have.

Mr. SCOTT. Thank you, Mr. Goodlatte.

Mr. Keller?

TESTIMONY OF THE HONORABLE RIC KELLER, A REPRESENTATIVE IN CONGRESS FROM THE STATE OF FLORIDA

Mr. KELLER. Thank you, Mr. Chairman and Mr. Ranking Member Forbes. It is also good to see my colleagues, Congressmen Gohmert and Coble and Lungren.

Aiming a laser beam into the cockpit of an airplane is a clear and present danger to the safety of all of those onboard the aircraft. This legislation is simple and straightforward. It makes it illegal to knowingly aim a laser pointer at an aircraft. Those who intentionally engage in such misconduct shall be fined or imprisoned not more than 5 years or both in the discretion of the judge.

This legislation was unanimously approved by all Democrats and Republicans on the House Judiciary Committee in the last Congress. It was then approved by the full House on a voice vote and the Senate also approved the legislation by unanimous consent after slightly amending the legislation to provide for limited exceptions by the Department of Defense and FAA, which we have included this time.

The problems caused by laser beam pranksters are more widespread than one might think. According to the Congressional Research Service and the Federal Aviation Administration, there have been over 500 incidences reported since 1990 where pilots have been disoriented or temporarily blinded by laser exposure. There have been 90 incidences in 2005 alone, according to the FAA.

These easily available laser pointers, like the one I purchased here for $12 at Staples earlier today, have enough power to cause vision problems in pilots from a distance of two miles.

I will demonstrate by looking at Mr. Sensenbrenner's portrait, and if you look at his face, you will see a laser beam right there, which has enough power to go almost two miles. I do this not merely for illustration purposes, but just because it is fun to poke fun at Mr. Sensenbrenner. I assume I will be subject to some sort of a voodoo-type penalty later.

But it is only a matter of time before one of these laser beam pranksters ends up killing over 200 people in a commercial airline crash. Surprisingly, there is currently no Federal statute on the books making it illegal to shine a laser beam into an aircraft's cockpit unless one attempts to use the Patriot Act to claim that an action was a terrorist act or other act of violence against a mass transportation system.

So far, none of the more than 500 incidents involving flight crew exposure to lasers have been linked to terrorism. Rather, it is often the case of pranksters making stupid choices to put pilots and their passengers at risk of dying. It is imperative that we send a message to the public that flight security is a serious issue. These acts of mischief will not be tolerated.

I wanted to learn what it was like to be inside an aircraft cockpit hit by a laser beam, so I spoke with Lieutenant Barry Smith from my hometown of Orlando, Florida, who was actually in the cockpit of a helicopter that was hit by a laser beam.

Lieutenant Smith is with the Seminole County Sheriff's Office. He and his partner were in a police helicopter searching for burglary suspects at night in a suburb of Orlando when a red laser beam hit the aircraft twice. Lieutenant Smith said the Plexiglas windshield of the helicopter spread out the light to be the size of a basketball. It shocked them. They were flying near a large tower with a red light and they mistakenly thought they may have flown too close to the tower, so they jerked the helicopter back and they became disoriented.

That is when they realized that they weren't near the tower at all. Then Lieutenant Smith began to worry that the light could have come from a laser site on a rifle. He wondered if they were about to be shot out of the sky. He told me, "It scared the heck out of us." In reality, it was a 31-year-old man with a small, pen-sized laser light standing in his backyard.

Currently, a handful of State legislatures, including Florida, have taken steps to address this matter. Governor Bush signed a bill into law making it illegal to focus the beam of a laser light at an aircraft. However, Federal legislation is needed because aircrafts travel across State lines and airports such as Ronald Reagan National Airport are located near State borders.

This legislation before us is needed to ensure the safety of pilots and passengers in all situations.

I also want to recognize and thank Congresswoman Waters and Congressman Johnson for showing up to this hearing. I didn't see you earlier. I am pleased that so many of you have come out to listen to what we have to say, and I appreciate your support in the past of this legislation and hopefully we can support it again tomorrow.

I yield back.

[The prepared statement of Mr. Keller follows:]

PREPARED STATEMENT OF THE HONORABLE RIC KELLER, A REPRESENTATIVE IN CONGRESS FROM THE STATE OF FLORIDA

Aiming a laser beam into the cockpit of an airplane is a clear and present danger to the safety of all those onboard the aircraft.

This legislation is simple and straightforward. It makes it illegal to knowingly aim a laser pointer at an aircraft. Those who intentionally engage in such misconduct, shall be fined or imprisoned not more than five years, or both, in the discretion of the judge.

This legislation was unanimously approved by all Democrats and Republicans on the House Judiciary Committee in the last Congress. It then was approved by the full House on a voice vote, and the Senate also approved the legislation by unanimous consent, after slightly amending the legislation to provide for limited exceptions by the Department of Defense and FAA.

The problems caused by laser beam pranksters are more widespread than one might think. According to the Congressional Research Service and the Federal Aviation Administration, there have been over 500 incidents reported since 1990 where pilots have been disoriented or temporarily blinded by laser exposure.

These easily available pen-sized laser pointers, like the one I purchased here for $12 at the House of Representatives Office Supply Store, have enough power to cause vision problems in pilots from a distance of two miles.

It's only a matter of time before one of these laser beam pranksters ends up killing over 200 people in a commercial airline crash.

Surprisingly, there is currently no federal statute on the books making it illegal to shine a laser beam into an aircraft's cockpit, unless one attempts to use the Patriot Act to claim that the action was a "terrorist attack or other attack of violence against a mass transportation system."

So far, none of the more than 500 incidents involving flight crew exposure to lasers have been linked to terrorism. Rather, it's often a case of pranksters making stupid choices to put pilots and their passengers at risk of dying. It is imperative that we send a message to the public that flight security is a serious issue. These acts of mischief will not be tolerated.

I wanted to learn what it was like to be in an aircraft cockpit hit by a laser beam, so I spoke with Lieutenant Barry Smith from my hometown of Orlando, Florida, who was actually in the cockpit of a helicopter that was hit with a laser beam.

Lieutenant Smith is with the Seminole County Sheriff's Office. He and his partner were in a police helicopter searching for burglary suspects at night in a suburb of Orlando, when a red laser beam hit the aircraft twice. Lieutenant Smith said the Plexiglas windshield of the helicopter spread out the light to be the size of a basketball. It shocked them. They were flying near a large tower with a red light, and they mistakenly thought they may have flown too close to the tower. They were disoriented and they immediately jerked the helicopter back.

When they realized that they weren't near the tower, Lieutenant Smith began to worry that the light could have come from a laser site on a rifle. He wondered if they were about to be shot out of they sky? He told me, "It scared the heck out of us."

In reality, it was a 31-year-old man, with a small, pen-sized laser light, standing in his yard.

Currently, a handful of state legislatures, including Florida's, are taking appropriate steps to address this matter. For example, on June 8, 2005, Governor Jeb Bush of Florida signed into law a bill making it illegal for any person to focus the beam of a laser lighting device at an aircraft. However, federal legislation is needed because aircrafts cross state lines and airports such as Ronald Reagan National Airport are located near state borders.

Clearly, this legislation before us is needed to ensure the safety of pilots and passengers in all situations, and I urge my colleagues to vote "Yes" on the legislation.

Mr. SCOTT. Thank you very much, Mr. Keller.

And I thank all of our witnesses.

At this time we will respond to questions. And I will reserve questions at this point and yield to my colleague from Virginia.

Mr. FORBES. Mr. Chairman, I think these three experts have exhausted this area so much, I have no additional questions.

But just, once again, thank them for their hard work in these areas.

Mr. GOODLATTE. If the gentleman will yield, except that the gentleman from Florida hasn't been bipartisan, and I don't believe he would get any objection from the Democratic side if he did the same thing with Former Chairman Brooks.

Ms. LOFGREN. I was going to ask for the laser so we could do that, Mr. Chairman.

Mr. SCOTT. The gentleman yields back.

The gentlelady from California?

Ms. WATERS. Mr. Chairman, I think that it is pretty straightforward and it certainly is well-understood. I just have no questions, and I am very supportive, and I would just ask that we just move with it.

Mr. SCOTT. Thank you very much.

The gentleman from Texas, Mr. Gohmert?

Mr. GOHMERT. Thank you, Mr. Chairman. And I might scare you and may make you want to rethink your positions, but I am in agreement with you.

But I do have a couple of questions for Mr. Goodlatte and Ms. Lofgren.

On the bill, I don't have Section 1030. Do you know offhand what the definition of "protected computer" is?

Because obviously that is an extremely important or integral part of this. It is referenced in Section 1030 and I apologize for not already having that, but I thought maybe if you had it handy.

Mr. GOODLATTE. I don't have it right in front of me, but I am reliably informed that it is a very broad definition of protected—it is almost any computer.

Mr. GOHMERT. Well, we can check on that. We want to make sure that it doesn't require some specific type of anti-spyware or something——

Mr. GOODLATTE. No, no, nothing like that.

Ms. LOFGREN. No, no, no.

Mr. GOHMERT.—in order to be protected.

Ms. LOFGREN. If you are connected.

Mr. GOHMERT. And then I am curious, in subsection C of 1030-A, "No person may bring a civil action under the laws of any State if such action is premised in whole or in part upon the defendants violating this section."

And I am just curious, I like the bill, but I am curious about the purpose of adding a civil bar to litigation.

Ms. LOFGREN. Here is the intent on that. There was concern that we were creating a litigation bonanza in 50 States. And there are State laws, you know, if there is an alleged violation of the State law, this doesn't do anything about that. You can bring an action under State law just as you can today.

But we didn't want to create a State law action premised on this Federal law, and that is basically what we said in this section. You can bring a Federal action under the Federal law, but we don't want to create a new State cause of action under the Federal law.

Mr. GOHMERT. So, that was my question. Is this also going to be able to be interpreted as barring any Federal action in Federal court?

Mr. GOODLATTE. No.

Ms. LOFGREN. No.

Mr. GOHMERT. Because I know people in America don't like frivolous lawsuits and that kind of thing, don't like to clutter the courts, but it sure seems like if people that are invading peoples' computers and their privacy were subject to civil liability of some kind in somebody's court, that it might have a deterrent effect.

Ms. LOFGREN. The point of this is to bring Federal action in Federal court under Federal law.

Mr. GOHMERT. Okay.

Ms. LOFGREN. And I think this section actually accomplishes that.

Mr. GOHMERT. Okay. That sounds good. And I will check on the definition in 1030 of protected computer.

I appreciate all of your work in getting this to this point. Thank you very much.

I yield back.

Mr. SCOTT. Thank you.

The gentleman from Georgia?

Mr. JOHNSON. Thank you, Mr. Chairman.

I have no questions. I would just simply like to say that the legislation appears to be forthright and directed toward issues that need to be addressed, and I support both bills 100 percent.

Thank you.

Mr. SCOTT. Thank you.

The gentleman from North Carolina?

Mr. COBLE. Thank you, Mr. Chairman.

As the distinguished Ranking Member, your colleague from Virginia, said when he conferred expert witness status upon our three witnesses, I remember, Mr. Chairman, that when you start examining expert witnesses, the examiner may end up looking foolish. So I will not assume that risk.

And I yield back.

Good to have you all with us, by the way.

Mr. SCOTT. Thank you.

The gentleman from California?

Mr. LUNGREN. Thank you very much, Mr. Chairman.

I just wanted to ask Ms. Lofgren and Mr. Goodlatte this, and that is much of the time when we deal with legislation in the area of rapidly expanding and changing technology, the technology outstrips our attempt to try and put reasonable regulation on it.

How have you tried to deal with that issue here? In other words, is our definition of spyware and phishing or activities that are similar to that inclusive enough such that we won't be able to—we will be here in another year or two trying to redo it because the technology has outstripped our effort to try to get at what we all agree is a practice that ought to be dealt with severely?

Mr. GOODLATTE. If I might, Mr. Lungren, that is a very good point, and it is in fact the hallmark of this legislation.

There is another version of legislation going through another Committee that addresses spyware from a much more regulatory approach, and while there are many commendable things in that legislation, one of the things that concerns us is that we could have the effect of stifling technology and being frankly out of date before it is even put into effect, unless we just go after the action and the intent, which is what this legislation does.

It goes after the bad actors. The definition of spyware is one that I think is a very encompassing one. Phishing and pharming are much more specific activities, but they are an ongoing problem on the Internet and they, plus the broader definition of spyware, I think would give this legislation, if it became law, give law enforcement the ability to go after people who have committed crimes without stifling technology on the Internet or without the problem of the technology moving on, as you say, and leaving this legislation irrelevant.

Ms. LOFGREN. I would just add, on page 2 of the bill, starting at line 9, it really defines the conduct, and it is really intend to engage in various fraud that is being prohibited in the computer environment. And we have done that, as Mr. Goodlatte has said, for a reason, not to get into regulation of existing technology not understanding what might be next.

As the gentleman knows, I represent Silicon Valley in the Congress and the technology community, at least as they have expressed it to me, much prefers this approach to the heavy regulatory approach that is being pursued in another Committee.

Mr. LUNGREN. And a second question, and that is: Some would say there are legitimate commercial uses of someone collecting information. If I go to a Web site repeatedly to purchase something, for instance, the Web site entity, the company that owns the Web site, might collect information about my buying preferences. Now, I may not like that, but I don't think that rises to the level of a crime.

Are we making sure that we differentiate between that and this kind of activity?

Ms. LOFGREN. Certainly. I mean, cookies are not phishing or pharming, and they actually, I mean, there are things you can do if you don't want to have cookies logged into your computer memory.

But if you look again to the definition, there is an intent definition to defraud, injure, cause damage, to impair the security protection, that really don't relate to the technology provisions that essentially allow the Web to function in its——

Mr. LUNGREN. And that is your intent, very much, in this legislation, to differentiate between that.

Ms. LOFGREN. That is correct.

Mr. LUNGREN. Thank you very much, Mr. Chairman.

Mr. SCOTT. Thank you. The gentleman yields back.

If there are no further questions, we will adjourn the hearing.

[Whereupon, at 1:35 p.m., the Subcommittee was adjourned.]

APPENDIX

MATERIAL SUBMITTED FOR THE HEARING RECORD

PREPARED STATEMENT OF THE HONORABLE SHEILA JACKSON LEE, A REPRESENTATIVE IN CONGRESS FROM THE STATE OF TEXAS, AND MEMBER, SUBCOMMITTEE ON CRIME, TERRORISM, AND HOMELAND SECURITY, ON H.R. 1525, THE "INTERNET SPYWARE (I-SPY) PREVENTION ACT OF 2007"

SHEILA JACKSON LEE
18TH DISTRICT, TEXAS

WASHINGTON OFFICE:
2435 Rayburn House Office Building
Washington, DC 20515
(202) 225-3816

DISTRICT OFFICE:
1919 SMITH STREET, SUITE 1180
THE GEORGE "MICKEY" LELAND FEDERAL BUILDING
HOUSTON, TX 77002
(713) 655-0050

ACRES HOME OFFICE:
6719 WEST MONTGOMERY, SUITE 204
HOUSTON, TX 77019
(713) 691-4882

HEIGHTS OFFICE:
420 WEST 19TH STREET
HOUSTON, TX 77008
(713) 861-4070

FIFTH WARD OFFICE:
3300 LYONS AVENUE, SUITE 301
HOUSTON, TX 77020

Congress of the United States
House of Representatives
Washington, DC 20515

COMMITTEES:
JUDICIARY
SUBCOMMITTEES:
COURTS, THE INTERNET, AND INTELLECTUAL PROPERTY
IMMIGRATION, CITIZENSHIP, REFUGEES, BORDER SECURITY, AND INTERNATIONAL LAW
CRIME, TERRORISM AND HOMELAND SECURITY

HOMELAND SECURITY
SUBCOMMITTEES:
CHAIR
TRANSPORTATION SECURITY AND INFRASTRUCTURE PROTECTION
BORDER, MARITIME, AND GLOBAL COUNTERTERRORISM

FOREIGN AFFAIRS
SUBCOMMITTEES:
AFRICA AND GLOBAL HEALTH
MIDDLE EAST AND SOUTH ASIA

SENIOR WHIP
DEMOCRATIC CAUCUS

WHIP
CONGRESSIONAL BLACK CAUCUS

CHAIR
CONGRESSIONAL CHILDREN'S CAUCUS

CONGRESSWOMAN SHEILA JACKSON LEE, OF TEXAS

STATEMENT BEFORE THE

JUDICIARY SUBCOMMITTEE ON
CRIME, TERRORISM, AND HOMELAND SECURITY

LEGISLATIVE HEARING: H.R. 1525
"INTERNET SPYWARE (I-SPY) PREVENTION ACT OF 2007"

MAY 1, 2007

Mr. Chairman, as a proud original co-sponsor of the legislation before us, I speak in strong support of H.R. 1525, the "Internet Spyware (I-SPY) Prevention Act of 2007." I wish to express my appreciation to my colleagues, Congresswoman Lofgren and Congressman Goodlatte for their tenacity in seeing this important legislation through.

H.R. 1525 amends the federal computer fraud and abuse statute to make it unlawful to access a computer without authorization or to intentionally exceed authorized access by causing a computer program or code to be copied onto the computer and using that program or code to transmit or obtain personal information (for example, first and last names, addresses, e-mail addresses, telephone numbers, Social Security numbers, drivers license numbers, or bank or credit account numbers).

Further, H.R. 744 discourages the practice of 'phishing.' As we all know too well, spyware is quickly becoming one of the biggest threats to consumers on the information superhighway. Spyware encompasses several potential risks including the promotion of identity theft by harvesting personal information from consumer's computers. Additionally, it can adversely affect businesses, as they are forced to sustain costs to block and remove spyware from employees' computers, in addition to the potential impact on productivity.

Spyware has been defined as "software that aids in gathering information about a person or organization without their knowledge and which may send such information to another entity with the consumer's consent, or asserts control over a computer with the consumer's knowledge." Among other things, criminals can use spyware to track every keystroke an individual makes, including credit card and social security numbers.

Some estimates suggest 25% of all personal computers contain some kind of spyware while other estimates show that spyware afflicts as many as 80-90% of all personal computers. Businesses are reporting several negative effects of spyware. Microsoft says evidence shows that sypware is "at least partially responsible for approximately one-half of all application crashes" reported to them, resulting in millions of dollars of unnecessary support calls.

Mr. Chairman, again, I am strongly in support of the legislation, and I yield back.

PREPARED STATEMENT OF THE HONORABLE SHEILA JACKSON LEE, A REPRESENTATIVE IN CONGRESS FROM THE STATE OF TEXAS, AND MEMBER, SUBCOMMITTEE ON CRIME, TERRORISM, AND HOMELAND SECURITY, ON H.R. 1615, THE "SECURING AIRCRAFT COCKPITS AGAINST LASERS ACT OF 2007"

SHEILA JACKSON LEE
18TH DISTRICT, TEXAS

WASHINGTON OFFICE:
2435 Rayburn House Office Building
Washington, DC 20515
(202) 225-3816

DISTRICT OFFICE:
1919 SMITH STREET, SUITE 1180
THE GEORGE "MICKEY" LELAND FEDERAL BUILDING
HOUSTON, TX 77002
(713) 655-0050

ACRES HOME OFFICE:
6719 WEST MONTGOMERY, SUITE 204
HOUSTON, TX 77019
(713) 691-4882

HEIGHTS OFFICE:
420 WEST 19TH STREET
HOUSTON, TX 77008
(713) 861-4070

FIFTH WARD OFFICE:
3300 LYONS AVENUE, SUITE 301
HOUSTON, TX 77020

Congress of the United States
House of Representatives
Washington, DC 20515

COMMITTEES:
JUDICIARY
SUBCOMMITTEES:
COURTS, THE INTERNET, AND INTELLECTUAL PROPERTY
IMMIGRATION, CITIZENSHIP, REFUGEES, BORDER SECURITY, AND INTERNATIONAL LAW
CRIME, TERRORISM AND HOMELAND SECURITY

HOMELAND SECURITY
SUBCOMMITTEES:
CHAIR
TRANSPORTATION SECURITY AND INFRASTRUCTURE PROTECTION
BORDER, MARITIME, AND GLOBAL COUNTERTERRORISM

FOREIGN AFFAIRS
SUBCOMMITTEES:
AFRICA AND GLOBAL HEALTH
MIDDLE EAST AND SOUTH ASIA

SENIOR WHIP
DEMOCRATIC CAUCUS

WHIP
CONGRESSIONAL BLACK CAUCUS

CHAIR
CONGRESSIONAL CHILDREN'S CAUCUS

CONGRESSWOMAN SHEILA JACKSON LEE, OF TEXAS

STATEMENT BEFORE THE

JUDICIARY SUBCOMMITTEE ON
CRIME, TERRORISM, AND HOMELAND SECURITY

LEGISLATIVE HEARING: H.R. 1615
"SECURING AIRCRAFT COCKPITS
AGAINST LASERS ACT OF 2007"

MAY 1, 2007

I thank the Chairman and Ranking Member for their effort and time in holding this legislative hearing and markup of H.R. 1615, the "Securing

Aircraft Cockpits Against Lasers Act of 2007." While the goal of this legislation – purportedly to keep our air passengers safe and to effect better "homeland security," I must point out my concern that this penal legislation be tailored as narrowly as possible to exclude only the evil sought to be prohibited.

Again, I understand that since 1990, there have been more than 400 incidents proscribed in this bill and more than 100 since November 2004.[1] Furthermore, I understand the relative threat that the act proscribed in this bill poses for pilots - FAA research has shown that laser illuminations can temporarily disorient or disable a pilot during critical stages of flight such as landing or take-off, and in some cases, may cause permanent damage.[2] This legislation *could* provide the legislative fix. However, there remains an element of vagueness in its provisions.

Section 2 references a "laser pointer," however, nowhere in the bill is the term described. Before we legislate criminal offenses that carry

---

[1] Statement of Nicholas A. Sabatini, Associate Administrator for Aviation Safety, Federal Aviation Administration, Before the Subcommittee on Aviation, Committee on Transportation and Infrastructure, U.S. House of Representatives, on Recent Laser Incidents and the Potential Impact on Aviation Safety. March 15, 2005.

[2] Van B. Nakagawara and Ronald W. Montgomery, "Laser Pointers: Their Potential Affects on Vision and Aviation Safety," DOT/FAA/AM-01/7, April 2001.

substantial penalties, we must take special care to use the most clear and narrowly-tailored terms to do so.

Mr. Chairman, for the above reasons, I have reservations regarding the wisdom of this legislation and look forward to hearing from the witness information that may assuage my concern.

Thank you, Mr. Chairman. I yield back the remainder of my time.

I

110TH CONGRESS
1ST SESSION

H. R. 1525

To amend title 18, United States Code, to discourage spyware, and for other purposes.

IN THE HOUSE OF REPRESENTATIVES

MARCH 14, 2007

Ms. ZOE LOFGREN of California (for herself, Mr. GOODLATTE, Ms. LINDA T. SÁNCHEZ of California, Mr. SMITH of Texas, and Ms. JACKSON-LEE of Texas) introduced the following bill; which was referred to the Committee on the Judiciary

---

A BILL

To amend title 18, United States Code, to discourage spyware, and for other purposes.

Be it enacted by the Senate and House of Representatives of the United States of America in Congress assembled,

SECTION 1. SHORT TITLE.

This Act may be cited as the "Internet Spyware (I–SPY) Prevention Act of 2007".

SEC. 2. PENALTIES FOR CERTAIN UNAUTHORIZED ACTIVITIES RELATING TO COMPUTERS.

(a) IN GENERAL.—Chapter 47 of title 18, is amended by inserting after section 1030 the following:

"§ 1030A. Illicit indirect use of protected computers

"(a) Whoever intentionally accesses a protected computer without authorization, or exceeds authorized access to a protected computer, by causing a computer program or code to be copied onto the protected computer, and intentionally uses that program or code in furtherance of another Federal criminal offense shall be fined under this title or imprisoned not more than 5 years, or both.

"(b) Whoever intentionally accesses a protected computer without authorization, or exceeds authorized access to a protected computer, by causing a computer program or code to be copied onto the protected computer, and by means of that program or code—

"(1) intentionally obtains, or transmits to another, personal information with the intent to defraud or injure a person or cause damage to a protected computer; or

"(2) intentionally impairs the security protection of the protected computer with the intent to defraud or injure a person or damage a protected computer;

shall be fined under this title or imprisoned not more than 2 years, or both.

"(c) No person may bring a civil action under the law of any State if such action is premised in whole or in part upon the defendant's violating this section. For the purposes of this subsection, the term 'State' includes the District of Columbia, Puerto Rico, and any other territory or possession of the United States.

"(d) As used in this section—

"(1) the terms 'protected computer' and 'exceeds authorized access' have, respectively, the meanings given those terms in section 1030; and

"(2) the term 'personal information' means—

"(A) a first and last name;

"(B) a home or other physical address, including street name;

"(C) an electronic mail address;

"(D) a telephone number;

"(E) a Social Security number, tax identification number, drivers license number, passport number, or any other government-issued identification number; or

"(F) a credit card or bank account number or any password or access code associated with a credit card or bank account.

"(e) This section does not prohibit any lawfully authorized investigative, protective, or intelligence activity of a law enforcement agency of the United States, a State, or a political subdivision of a State, or of an intelligence agency of the United States.".

(b) CONFORMING AMENDMENT.—The table of sections at the beginning of chapter 47 of title 18, is amended by inserting after the item relating to section 1030 the following new item:

"1030A. Illicit indirect use of protected computers.".

SEC. 3. AUTHORIZATION OF APPROPRIATIONS.

In addition to any other sums otherwise authorized to be appropriated for this purpose, there are authorized to be appropriated for each of fiscal years 2008 through 2011, the sum of $10,000,000 to the Attorney General for prosecutions needed to discourage the use of spyware and the practices commonly called phishing and pharming.

SEC. 4. FINDINGS AND SENSE OF CONGRESS CONCERNING THE ENFORCEMENT OF CERTAIN CYBERCRIMES.

(a) FINDINGS.—Congress makes the following findings:

(1) Software and electronic communications are increasingly being used by criminals to invade individuals' and businesses' computers without authorization.

(2) Two particularly egregious types of such schemes are the use of spyware and phishing scams.

(3) These schemes are often used to obtain personal information, such as bank account and credit

card numbers, which can then be used as a means to commit other types of theft.

(4) In addition to the devastating damage that these heinous activities can inflict on individuals and businesses, they also undermine the confidence that citizens have in using the Internet.

(5) The continued development of innovative technologies in response to consumer demand is crucial in the fight against spyware.

(b) SENSE OF CONGRESS.—Because of the serious nature of these offenses, and the Internet's unique importance in the daily lives of citizens and in interstate commerce, it is the sense of Congress that the Department of Justice should use the amendments made by this Act, and all other available tools, vigorously to prosecute those who use spyware to commit crimes and those that conduct phishing and pharming scams.

○

I

110TH CONGRESS
1ST SESSION

H. R. 1615

To amend title 18, United States Code, to provide penalties for aiming laser pointers at airplanes, and for other purposes.

IN THE HOUSE OF REPRESENTATIVES

MARCH 21, 2007

Mr. KELLER of Florida (for himself and Mr. FORBES) introduced the following bill; which was referred to the Committee on the Judiciary

---

A BILL

To amend title 18, United States Code, to provide penalties for aiming laser pointers at airplanes, and for other purposes.

Be it enacted by the Senate and House of Representatives of the United States of America in Congress assembled,

SECTION 1. SHORT TITLE.

This Act may be cited as the "Securing Aircraft Cockpits Against Lasers Act of 2007".

SEC. 2. PROHIBITION AGAINST AIMING A LASER POINTER AT AN AIRCRAFT.

(a) OFFENSE.—Chapter 2 of title 18, United States Code, is amended by adding at the end the following:

“§ 39A. Aiming a laser pointer at an aircraft

“(a) Whoever knowingly aims the beam of a laser pointer at an aircraft in the special aircraft jurisdiction of the United States, or at the flight path of such an aircraft, shall be fined under this title or imprisoned not more than 5 years, or both.

“(b) As used in this section, the term ‘laser pointer’ means any device designed or used to amplify electromagnetic radiation by stimulated emission that emits a beam designed to be used by the operator as a pointer or highlighter to indicate, mark, or identify a specific position, place, item, or object.

“(c) This section does not prohibit aiming a beam of a laser pointer at an aircraft, or the flight path of such an aircraft, by—

“(1) an authorized individual in the conduct of research and development or flight test operations conducted by an aircraft manufacturer, the Federal Aviation Administration, or any other person authorized by the Federal Aviation Administration to conduct such research and development or flight test operations;

“(2) members or elements of the Department of Defense or Department of Homeland Security acting in an official capacity for the purpose of research, development, operations, testing or training; or

"(3) by an individual using a laser emergency signaling device to send an emergency distress signal.

"(d) The Attorney General, in consultation with the Secretary of Transportation, may provide by regulation, after public notice and comment, such additional exceptions to this section, as may be necessary and appropriate. The Attorney General shall provide written notification of any proposed regulations under this section to the Committees on the Judiciary of the House and Senate, the Committee on Transportation and Infrastructure in the House, and the Committee on Commerce, Science and Transportation in the Senate not less than 90 days before such regulations become final.".

(b) AMENDMENT TO TABLE OF SECTIONS.—The table of sections at the beginning of chapter 2 of title 18, United States Code, is amended by adding at the end the following new item:

"39A. Aiming a laser pointer at an aircraft.".

○

APEX
TP
9780160817045
76901953-62